The New Northwest

The New Northwest

The Photographs of the Frank Crean Expeditions 1908–1909

BILL WAISER

Front cover painting: the original cover from Frank Crean's "Report of Exploration, season of 1908–1909," reproduced courtesy *The Beaver* magazine
Back cover photograph: Métis boy and girl at Île-à-la-Crosse, 1908. Courtesy Saskatchewan Archives Board
Cover design by NEXT Communications Inc.

The publisher gratefully acknowledges the support received from The Canada Council, Communications Canada, the Saskatchewan Arts Board, and the University of Saskatchewan.

Printed and bound in Canada by Best/Gagné Book Manufacturers
93 94 95 96 97 / 5 4 3 2 1

Canadian Cataloguing in Publication Data
Waiser, W. A.

The new northwest
ISBN 1–895618–22–3

1. Crean, Frank J. P. (Frank Joseph Patrick), 1875– - Journeys - Saskatchewan, Northern. 2. Crean, Frank J. P. (Frank Joseph Patrick), 1875– - Journeys - Alberta, Northern. 3. Saskatchewan, Northern - Discovery and exploration. 4. Saskatchewan, Northern - Pictorial works. 5. Alberta, Northern - Discovery and exploration. 6. Alberta, Northern - Pictorial works. I. Title.

FC3205.3.W34 1993 917.1204'2 C93–098112–X
F1060.9.W34 1993

FIFTH HOUSE LTD.
620 Duchess Street
Saskatoon, SK S7K 0R1

Contents

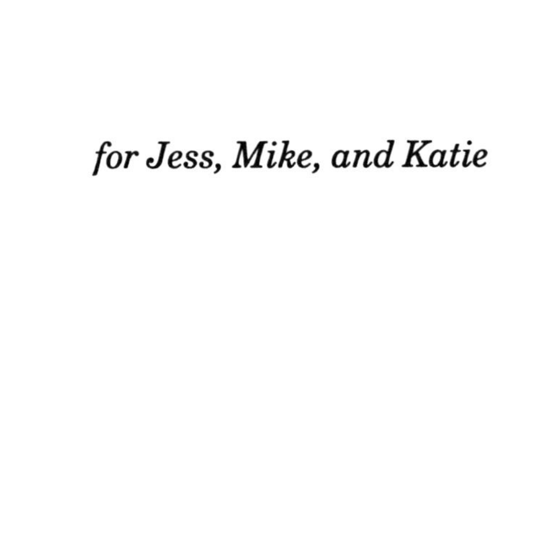

for Jess, Mike, and Katie

Preface

In the early twentieth century, the Canadian government became concerned about the rapid rate at which homestead lands were being occupied in western Canada and was worried that prospective immigrants would settle elsewhere if new lands were not made available. This apparent need for a new agricultural frontier rekindled interest in the potential of Canada's northland and in 1908–09, the federal Department of the Interior sent civil engineer Frank Crean to assess the agricultural capabilities of the boreal forest region north of Prince Albert, Saskatchewan. These "New Northwest" expeditions were the first of their kind to the middle north and created quite a sensation; not only did Crean predict a great mixed-farming future for the region, but he also captured on film one of the lesser known parts of Canada at the turn of the century. *The New Northwest* is the story of this fascination—albeit shortlived—with the agricultural potential of Canada's northland, and of the photographs of the people and places in northern Saskatchewan before World War One.

The existence of the Crean photographs was originally brought to my attention by Walter Hildebrandt. The Saskatchewan Archives Board, in particular D'Arcy Hande, director of the Saskatoon office, attended to my many requests regarding the Crean collection, and this kind and professional assistance is gratefully acknowledged. Jeff Murray of the National Archives of Canada checked a number of possible leads in his usual efficient manner. Anne Morton of the Hudson's Bay Company Archives answered a number of queries about company personnel. Frank Crean's great nephew, Richard Crean of Toronto, provided information on the Crean family and took a continuing interest in the project. My colleagues, Jim Miller and Ted Regehr, read and expertly commented on the draft manuscript; their advice and guidance over the years has been greatly valued. George Duff turned a list of places and features into two fine maps. Jean Horosko assisted with the preparation of the photographs and helped keep things in perspective.

Funding for the research and preparation (including photographs and maps) of *The New Northwest* was generously provided by several University of Saskatchewan sources: the President's SSHRC Fund, the College of Arts and Science, and the University Publications Fund. Without this financial assistance, the project would still be incomplete.

That the story of the New Northwest is available in print is a credit to Fifth House and its abiding interest in the region and its history. A special thank you is owed to Nora Russell, who guided the manuscript through its various stages with her usual care and artistic sensitivity.

Finally, to appreciate the concept of the New Northwest, the reader might consider a passage from George Bernard Shaw's *Man and Superman.* When Don Juan asked directions, he was advised, "Oh, the frontier is only the difference between two ways of looking at things. Any road will take you across it if you really want to get there."

The New Northwest

The Photographs of the Frank Crean Expeditions 1908–1909

Frank Oliver had scarcely assumed his new duties as federal minister of the interior in 1905 when he received a troubling report from one of his staff members. For the past few years, Robert Young, in his capacity as superintendent of railway and swamp lands, had been monitoring the rate at which homesteads were being taken up in western Canada and reached the startling conclusion that the current rate of immigration—almost 150,000 in 1905 alone—would soon exhaust the available lands. It was a surprising finding, especially when the region was being touted world-wide as a land of unlimited opportunities, the so-called "last best west." For Young, the solution was simple. If Canada was to continue to attract tens of thousands of immigrant farmers annually, then the vast country north of the North Saskatchewan River would have to be opened up to settlement. To illustrate the great potential of Canada's northland, Young had a special map prepared to accompany his report that neatly summarized all the favourable information about the region.

Young's championing of the Northwest was nothing new. When Canada secured control of the vast territory west of Ontario in 1870, it was generally assumed that western settlement would be initially restricted to the prairie parkland or "fertile belt" along the Assiniboine and North Saskatchewan rivers. Another body of opinion, however, believed that the best agricultural land in the western interior was actually to be found in the boreal forest zone. Long-time western resident Bishop Alexandre Taché claimed that the capabilities of the fertile belt had been vastly overrated—it was better suited for grazing, not cultivation—and instead drew attention to the wonderful crops that had been raised at northern missions since the 1850s.[1] Captain William F. Butler, who

1 A.A. Taché, *Sketch of the North-West of America*, trans. (Montreal 1870), 10–12; 19; 28.

coined the phrase, "the great lone land," was also an ardent supporter of the northland. In the account of his journey up the mighty Peace River in the spring of 1872, he predicted, "It will yet be found that there are ten acres of fertile land lying north of the North Saskatchewan for every acre lying south of it."[2] This seemed to be the case as a number of exploratory surveys returned from the field in the 1870s with stories of the luxuriant gardens they had found at northern fur trade posts. One of the northland's greatest advocates at this time was the botanist John Macoun, who was astounded by the richness of the vegetation that he had encountered during two exploratory trips to the Peace River country in 1872 and 1875. He subsequently portrayed the region as a kind of agricultural Eden. "The vastness seems to overpower the mind," he confided to his journal at Fort Chipewyan on Lake Athabasca in 1875, "and cause that benumbing feeling which we are prone to feel when in the presence of something we cannot grasp."[3]

These glowing reports about the northland's potential caused Sandford Fleming, the civil engineer in charge of finding the best possible route for the Canadian Pacific Railway, to consider sending the rail line by a northern pass through the Rockies. In fact, throughout the 1870s, there had been considerable debate as to whether the main line should be sent via the favoured Yellowhead Pass or follow a more northwesterly route to the Pacific. When the new Canadian Pacific Syndicate assumed control of the project in the spring of 1881, however, it decided to build the main line directly west from Winnipeg across the southern prairie. This decision, one of the most controversial in western Canadian history, had a tremendous impact on the region's development, by focusing settlement activity for the next two decades along a thin line through the southern prairies. According to one observer, the Northwest effectively became the West.[4] Meanwhile, those communities throughout the parkland that anxiously awaited the coming of the railway—places like Prince Albert, Battleford, and Edmonton—were reduced overnight to economic backwaters. They found themselves two hundred miles north of the settlement frontier, a situation that bred a strong sense of grievance and betrayal.

Although the rerouting of the CPR rad-

---

2 W.F. Butler, *The Wild Northland* (London 1873), 358.

3 Quoted in W.A. Waiser, *The Field Naturalist: John Macoun, the Geological Survey, and Natural Science* (Toronto 1989), 30.

4 M. Zaslow, *The Opening of the Canadian North 1870–1914* (Toronto 1971), 29.

ically altered the shape of western development, the northland was not forgotten. While homesteads were rapidly marked out in the open prairie district, a number of special teams were sent into the territory north of the fertile belt to conduct general reconnaissance surveys. This exploratory work nicely complemented that of the Geological Survey of Canada, whose officers were largely responsible for gathering resource information on the more remote regions of the young dominion. In the late 1880s, for example, while geologist George Dawson toiled up the Yukon River, his colleague Robert Bell roamed the Hudson Bay lowlands. In each instance, the men returned with favourable reports about the district's potential.

The Northwest also continued to stir the imaginations of Canadian politicians. Even though settlement of the prairie west would preoccupy successive federal administrations well into the twentieth century, there remained a widespread belief that Canada had been blessed with millions of acres of northern land whose idle richness was just waiting to be exploited. John Christian Schultz, a Conservative senator from Manitoba and future lieutenant-governor of the province, consequently had little difficulty convincing his fellow senators in May 1887 to establish a select committee to examine the natural food products of the Northwest.[5] When Schultz tabled the committee's report the following month, he noted that its findings confirmed the popular belief that

John Christian Schultz was an ardent believer in the northland's potential and headed two Senate hearings on the region in the 1880s. (Provincial Archives of Manitoba N10487. John Christian Schultz c1888. Photograph by Harry Sproule, Ottawa)

5 Schultz had appeared before the first Senate investigation of the region seventeen years earlier. See CANADA. Senate, *Journals*, 1870, vol. 3, appendix 1, "Report of the Select Committee of the Senate on the subject of Rupert's Land, Red River, and the North-West Territory."

crop yields were greater near the northern limit of successful growth and that the land seemed "prepared by the hands of God for the homes of civilized man."[6]

The Senate went through a similar exercise less than a year later when the Manitoba senator proposed another round of committee hearings. This time, however, the motion was approved only after a spirited debate, thanks to Liberal Senator L.G. Power of Halifax, who questioned the need for gathering more information about the northland when there were still homesteads available in the prairie district. As Power bluntly summed up the situation, "It will keep . . . We have plenty of time before us." These remarks were something of a red flag to Schultz and his supporters, who believed that the greatest obstacle to the northland's development was a general ignorance of its vast resources. Senator M.-A. Girard, the former Manitoba premier, maintained that neither money nor time should be spared in making the region better known to the world. Henry Kaulbach, also a Nova Scotia senator, attributed a more sinister intent to Power's comments, accusing him of wanting to see the development of the north deliberately held back. Schultz, for his part, was personally offended. "I am proud and pleased to say that even in its darkest days," he declared in an oblique reference to his ignoble part in the Red River Resistance of 1869–70, "I have never lost faith in the home of my adoption."[7]

The second Schultz committee was concerned specifically with the commercial and agricultural resources of the great Mackenzie Basin—what Schultz termed Canada's "Great Reserve" north of the Saskatchewan watershed, east of the Rocky Mountains and west of Hudson Bay. And unlike the brief 1887 hearings, it made a concerted effort to gather as much information as possible; traders, missionaries, scientists, and politicians who had either lived in or visited the region were consequently called upon to give evidence, while a fifty-question survey was sent to 150 respondents. The image that emerged from the hearings was one of a super land that would be the home of super men. "It is a land perhaps not flowing with honey," one witness remarked, "but it is a land teeming with everything that makes the heart of man glad."[8] In the end, though, Power's words rang true. In tabling the

6 CANADA. Senate, *Journals*, 1887, vol. 21, appendix 1, "Report and Minutes of Evidence of the Select Committee of the Senate on the existing Natural Food Products of the North-West Territories, and the best means of conserving and increasing them," 6.

7 CANADA. Senate, *Debates,* 27 March 1888, 225.

8 CANADA. Senate, *Journals,* 1888, vol. 22, no. 1, "Report of the Select Committee of the Senate

committee's third and final report, Schultz reluctantly admitted that colonization of the region would have to wait until the prairies had been settled.

Throughout the 1890s, the northland continued to excite the Canadian imagination. It was a fashionable destination for North American adventurers, many of whom produced popular accounts, complete with photographs, of their travels through the region. It also continued to be probed by the Geological Survey of Canada, which suspected that some of the blank spaces on existing maps might provide valuable scientific clues that would help unravel the country's geological history. Joseph B. Tyrrell, for example, spent several consecutive field seasons exploring the boreal forest and neighbouring barren lands west of Hudson Bay, and subsequently concluded that the great glaciers that had once covered the prairies had originated in this region.[9] These kinds of findings, however, created less of a sensation than the valuable mineral deposits that were being found in the region. In fact, it appeared that the great predictions of the northland's hidden wealth were finally being realized when gold was discovered in the Yukon in 1896.

The Klondike Gold Rush—what one author has described as the most important single event in northern history[10]—cultivated, to the point of absurd exaggeration, the notion of the north as a vast, untapped hinterland of untold riches. Thousands of would-be prospectors from all walks of life quit their jobs and headed north in lemming-like fashion to join the ranks of Klondike Kings. The stampede also helped generate an unprecedented boom in the Canadian economy, which prompted Prime Minister Wilfrid Laurier to claim that the twentieth century would belong to Canada. It was a bold declaration, but one that perfectly captured the spirit of the age; immigrants were pouring into the country in record numbers, while two new transcontinental railways were under construction. It was not a time for pessimism, as one sorry government employee learned in 1904 when he was nearly dismissed for producing a negative report on the settlement prospects of the Peace River country.[11] There was a general feeling that anything

appointed to Inquire into the Resources of the Great Mackenzie Basin," 247.

9 Senator Schultz took a great interest in Tyrrell's northern work, and when he learned that limited government funds might prevent the geologist from continuing his exploratory work, he offered to make him a justice of the peace for the region in order to finance future trips. *Thomas Fisher Rare Book Library* [*TFRBL*], J.B. Tyrrell papers, Box 4, J.C. Schultz to J.B. Tyrrell, 2 April 1892.

10 Zaslow, *The Opening of the Canadian North*, 101.

11 W.A. Waiser, "A Bear Garden: James Melville Macoun and the 1904 Peace River Controversy,"

was possible and that the limits of Canada's growth were confined only by small thinking. As for those who had been heralding the virtues of the northland, the time had arrived, in poet Charles Mair's words, to "right the wronged land at last."[12]

It was around this time that Robert Evans Young assumed his position as superintendent of railway and swamp lands with the federal Department of the Interior. Young was no stranger to western Canada or its potential. Born in Georgetown, Ontario, in 1861, he had been educated at Albert College in Belleville, where he likely fell under the influence of Professor John Macoun, who was busy at the time remaking almost singlehandedly the image of the Northwest. Young passed his Dominion Lands Surveyor examinations in 1882, and except for a brief stint with the Winnipeg Rifles during the North-West Rebellion, spent the next two decades working in British Columbia and Manitoba, during which time he conducted a special survey of the city of Winnipeg. He thus had considerable practical experience when in March 1901 he was given the job of overseeing the disposition of outstanding western land grants to various railway companies and their subsidiaries. Since the signing of the CPR contract in 1881, the federal government had been offering huge western land subsidies to encourage and help fund railway construction. And by the turn of the century, almost 29 million acres had been earned by the railways through actual construction. This was more land than the three maritime provinces combined. The railways, however, had been extremely slow in selecting their land—only a little more than 2 million acres had been patented by 1898—so there were large reserves closed to settlement at the very time that western Canada had begun to attract homesteaders in record numbers. Under intense government pressure, particularly from Minister of the Interior Clifford Sifton, the railways finally began to cash in these western land credits to take advantage of the settlement boom. Young's new role as superintendent of railway lands was to facilitate this process, and by the time his position was made official less than three years later in February 1904, his department had issued patents for 13 million acres of railway lands.[13]

It was as a direct result of his work on the railway lands question that Young

---

Canadian Historical Review, 67, 1 (March 1986), 42–61.

12 C. Mair, "Open the Bay," from *Dreamland and Other Poems* (Toronto 1974), 170.

13 *National Archives of Canada* [*NAC*], Government Archives Division, Privy Council, RG 2, v. 869, f. 879D, v. U17, PC 331(a); C. Martin, *"Dominion Lands" Policy* (Toronto 1973), 96–99; D.J. Hall, *Clifford Sifton:*

began to wonder whether western Canada could continue to absorb the steady flow of settlers in the new century. Even though millions of acres of railway lands would now be available for sale, it appeared only a matter of time at the current rate of immigration before all the free homesteads in the western interior would be gone. Greatly troubled by this prospect, Young concluded that the Laurier government had to shift its attention northward, and in 1905 prepared a special report to this effect. Young's concerns would have received a sympathetic hearing from the new minister of the interior, Frank Oliver. The member of parliament for Alberta and outspoken editor of the Edmonton *Daily Bulletin*, Oliver had long been a proponent of the Peace River country and had made a written submission to the 1887 Schultz committee on the virtues of the region. The new minister's more immediate concern upon taking over from Sifton, however, was a major overhaul of dominion lands legislation, including the cancellation of all outstanding railway grants in favour of creating more homestead land.[14]

Saskatchewan Senator T.O. Davis saw Prince Albert serving as the gateway to the New Northwest. (Provincial Archives of Manitoba N14983. Thomas Osborne Davis 1901–02. From John A. Cooper, *Men of Canada,* p.169)

The Young report was not simply shelved. It also greatly interested T.O. Davis, a veteran Prince Albert merchant and new Liberal senator who was anxious to rekindle some of the earlier expectations for the region. Davis saw in the report a chance to promote Prince Albert as the base from which northern resources might be developed, and he consequently rose in his seat in the Senate on 24

The Young Napoleon 1861–1900 (Vancouver 1981), 254–55; D.J. Hall, *Clifford Sifton: The Lonely Eminence 1901–1929* (Vancouver 1985), 57–61.

14 Martin, *"Dominion Lands" Policy*, 97–99.

January 1907 and called for an investigation of the country north of the Saskatchewan watershed, east of the Rocky Mountains and west of Hudson Bay. In support of his motion, Davis recalled the fine work that had been done by the 1887 Schultz committee, but pointed out that much more was now known about the northland, especially with the discovery of gold in the Yukon. He also argued that it would be valuable to take stock of the resources in the northern hinterland of the new provinces of Saskatchewan and Alberta.

The Davis proposal met with general approval. In speaking for the government about the need for such an inquiry, the aged parliamentarian Sir Richard Cartwright remarked, "We have barely scratched the surface, and yet we have found such places . . . where there are enormous resources of which we never dreamed."[15] Others spoke of the urgent need to send exploratory parties to the region, while one senator remarked that the much maligned chamber would be seen to be doing something of great public value. Before the motion was adopted, however, it was agreed to broaden the scope of the investigation to include the districts of Keewatin and Ungava.

As in past hearings of this nature, a number of Geological Survey officers were summoned before the Davis committee to give expert testimony in the general areas of agriculture, forestry, fisheries, minerals, climate, settlement, and communication. Many of these men had spent years probing a particular district—whether it be the great Mackenzie Valley, the rugged Labrador peninsula, or the desolate barrens of Keewatin—and they not only spoke with some authority, but welcomed the opportunity to give their impressions of the region's potential. In fact, they tended to be as accommodating as possible. When, for example, Davis took issue with R.G. McConnell's disparaging remarks about the frequency of early frosts in the Peace River country, McConnell quickly corrected himself and suggested that the situation would be different once the land had been broken.[16] The committee also heard from several political figures who represented the regions under review. Not unexpectedly, these men all spoke in glowing terms about the northland, especially the mayor of Prince Albert, who confessed, "It is such an immense country it is hard to grasp the possibilities of it. You might travel over it for years and know very little about it."[17]

15 CANADA. Senate, *Debates*, 24 January 1907, 138.

16 E.J. Chambers, ed., *Canada's Fertile Northland: A Glimpse of the Enormous Resources of Part of the Unexplored Regions of the Dominion* (Ottawa 1908), 53.

Even Robert Young, the man whose report had initially sparked the Senate committee hearings, made a brief appearance; armed with a series of new maps and tables that confirmed his earlier warning that available homestead lands would soon be exhausted, he called for the immediate, systematic exploration of the northern reaches of the western provinces.

The Davis committee reported to the Senate in mid-April 1907, less than four months after the inquiry had been approved. Its findings paralleled those of the earlier Schultz committees; the more sensational evidence was neatly summarized and capped with the prediction that "[T]he great northland appears, at last, to be on the eve of exploitation."[18] Unlike the results of the earlier investigations, however, the committee's report and accompanying testimony were not simply buried in an appendix to the Senate *Journals* and quickly forgotten. From the outset of the inquiry, Senator Davis had expressed a desire to publish the findings in pamphlet format—a project now made possible by the unusual situation of the committee having completed its work with money to spare. The editing of the inquiry documents was consequently turned over to Ernest J. Chambers, gentleman usher of the black rod and a former journalist,[19] who was expected to turn the proceedings into a more palatable product for general consumption. His supervisor for the project was none other than Robert Young.

Canada's Fertile Northland was released in January 1908. The bulk of the pamphlet consisted of the committee evidence, which was reproduced, without comment, in its entirety. Chambers also furnished a brief introduction, which opened with the warning that Canada's future expansion would depend upon the development of her northland. That the resources of this region were vast yet little known was driven home by the pamphlet subtitle, "a glimpse of the enormous resources of part of the unexplored regions of the Dominion." The most provocative part of the pamphlet was the series of maps specially prepared under Young's direction. They featured summer isothermal lines in northern districts,[20] average

17 *Ibid.*, 80.

18 *Ibid.*, 1.

19 Chambers later served as chief press censor during the Great War. See J. Keshen, "All the News That Was Fit To Print: Ernest J. Chambers and Information Control in Canada," *Canadian Historical Review*, v. 73, n. 3, September 1992, 315–43.

20 Isothermal lines had been used since the 1850s to support the idea that the climate in the Northwest was suitable for agriculture. The fact that summer isotherms swept northwestward suggested that

possible hours of sunshine in the summer months, soil conditions, known mineral deposits, locations where wheat had been grown, and the distances between Hudson Bay and major European ports. One particular map featured the outline of the Siberian province of Tobolsk superimposed at its correct latitude over northwestern Canada; the accompanying legend explained that the province, whose southern boundary was roughly one hundred miles north of Edmonton, had a population of almost 1.5 million people and produced over 6 million bushels of wheat in 1900. It was a brilliant comparison, especially when it was pointed out that Canada had the added benefit of British institutions.

The distribution and promotion of *Canada's Fertile Northland* was carefully orchestrated by Young. Copies of the pamphlet were sent to all of Canada's major newspapers and magazines, which gave the story feature coverage. The Toronto *Daily Star,* for example, reported that the potential of the north was "a cause of amazement" and that Canada should no longer be simply regarded as a fringe of settlements along the American boundary. "We speak of the Great West of today," the paper observed. "To-morrow the land of wonder will be the Great North."[21] Young, in fact, was so pleased with the press reception of the pamphlet that he hurriedly arranged to have the reports of the earlier Schultz committees published under the title, *The Great Mackenzie Basin.*[22] As part of his promotional campaign, he also

The promotion of Canada's northland achieved its fullest expression during Frank Oliver's tenure as federal minister of the interior. (Provincial Archives of Alberta, Ernest Brown Collection B7106)

latitude was a faulty indicator of a region's potential. See G.S. Dunbar, "Isotherms and Politics: Perceptions of the Northwest in the 1850s," in A.W. Rasporich and H.C. Klassen, eds., *Prairie Perspectives 2* (Toronto 1973), 80–101.

21 Toronto *Daily Star*, 15 February 1908.

22 E.J. Chambers, ed., *The Great Mackenzie Basin: Reports of the Select Committees of the Senate* (Ottawa

made a series of guest lectures, taking along his maps, charts, and infectious enthusiasm for the region. The message was always the same: Canada was quickly running out of free homestead land, and the north, once dismissed as a frozen wasteland, represented the dominion's future salvation. It was there for the taking. The big question, Young told the Canadian Club of Toronto, was "whether the present and the coming generations of Canadians would be worthy of the great heritage which God had given them, and make the best of it for the greatest ends that He sought."[23]

Young's efforts paid great dividends. At his own suggestion, he was placed in charge of a new branch whose purpose was to collect all available information on the land between Hudson Bay and the Rocky Mountains—what was now being referred to as the New Northwest.[24] The Department of the Interior, meanwhile, was swamped with over ten thousand individual requests for *Canada's Fertile Northland*. And in a media blitz somewhat reminiscent of the heady days of the Klondike Gold Rush, newspapers regularly carried individual testimonials about the north and its great potential; anyone who lived in or travelled through the region was interviewed for his or her impressions.[25] Even Young's boss, the minister of the interior, got caught up in the excitement. During a debate over the amount of good land in western Canada, Oliver told the House, "But beyond that [the prairies] there is a stupendous area of which a great part may be suitable for settlement."[26]

The next logical step was the detailed investigation of the region. Throughout the Davis inquiry hearings, witnesses repeatedly emphasized that the resources of the northland were not only unlimited, but that the region was still essentially unknown. The committee consequently recommended that the Laurier government dispatch a team of exploratory parties to the region—a view shared by Young himself. He fully realized that development of the north would take place only after the land had been formally surveyed and the re-

1908).

23 "Canada's Fertile Northland," *The Christian Guardian*, 1 April 1908, 12.

24 *National Museums of Canada* [*NMC*], National Museum of Natural Sciences, Botany Division, John Macoun correspondence, R.E. Young to J.M. Macoun, 31 January 1908.

25 See, for example, "Seen Through a Surveyor's Glasses," Edmonton *Bulletin*, 5 December 1908; "Generation of Axe Work Before Canada," Montreal *Herald*, 12 January 1909; "Mineral Wealth in the North," Prince Albert *Times*, 28 July 1909.

26 CANADA. House of Commons, *Debates*, 23 June 1908, 11127.

sources evaluated. As he explained to the Commons Standing Committee on Agriculture and Colonization during his promotion of *Canada's Fertile Northland,* "there is a deduction that seems to follow from all these statements if they amount to anything at all . . . it is time that we knew more about the country."[27]

The region that Young targeted for detailed investigation was northcentral Saskatchewan. There were several apparent reasons for this choice. Since the 1870s, there had been a steady number of assessments of the agricultural potential of the Peace River country in northern Alberta; some of the most prominent government scientists of the time had done field work there. The same could not be said of northern Saskatchewan. Although the northern boundary of the new province was set at sixty degrees in 1905 because it was believed to be the northern limit of agriculture, there had been no specific survey of the region's agricultural promise. In fact, most expeditions to the area had confined their activities to the waterways and rarely ventured inland.[28] That this region might have great potential was evidenced by the spread of settlement along the southern edge of the boreal forest beginning around the turn of the century after the railway had finally reached Prince Albert. The early success of mixed farming in the area suggested that the land further to the north might also support agriculture; only detailed assessment would tell.[29] Finally, in sending an expedition into northern Saskatchewan, Young could count on the support of Senator Davis and whatever influence he had with the government. Although he got along well with his polit-

27 CANADA. House of Commons, *Journals*, 1907–08, vol. 43, part II, appendix 2, "Report of the Select Standing Committee on Agriculture and Colonization," 153.

28 In the mid-1890s, Joseph Burr Tyrrell of the Geological Survey spent several field seasons mapping the geology of northern Saskatchewan. William McInnes, another Survey officer, continued this work a decade later. See W.O. Kupsch, "Chronology of Prospecting, Exploration, and Mining in Northern Saskatchewan to 1985," in W.O. Kupsch and S.D. Hanson, eds., *Gold and other stories as told to Berry Richards* (Regina 1986), appendix c.

29 It was believed by some observers that the region would have special appeal to the recent wave of immigrants from continental Europe. J.M. Macoun of the Geological Survey of Canada told Young in February 1908: "For my own part I have never been surprised that people should go insane who live on the prairie, and one who has not thought about it cannot possibly realize the homesickness and loneliness of the poor unfortunates who come from well wooded and well watered districts in Europe to our own western prairies. These people will find just the kind of country that they are accustomed to in the wooded and partly wooded lands north of the Saskatchewan." *NMC,* National Museum of Natural Sciences, Botany Division, J.M. Macoun to R.E. Young, 21 February 1908.

ical masters, Young was still a civil servant and could only do so much on his own.

The man who was selected to head the Saskatchewan expedition was Francis Joseph Patrick Crean (pronounced Crane), a civil engineer and clerk in Young's division. Born in Dublin, Ireland, 6 March 1875, Frank was the seventh of eight children of Michael (and Emma) Crean, a barrister with the Irish Land Commission. He likely attended Belvedere College, but it is not clear where he secured his engineering training; he was not a member of the Irish or English surveyors societies. Crean served with the 1st Royal Dragoons, Roberts' Horse, in South Africa in 1899, but little else is known about his activities at this time. It is also uncertain when he emigrated to Canada or why—although he is vaguely remembered today by family members as the proverbial black sheep.[30] The earliest record of his Canadian activities is 1903, when he was cruising timber thirty-five miles north of Prince Albert for the Bell Brothers Lumber Company.[31] He also spent time in Winnipeg, for he claimed to be a personal friend of C.C. Chipman, Canadian commissioner of the Hudson's Bay Company, and would later marry the daughter of a former member of

Frank Crean in the field, 1909. (Saskatchewan Archives Board S-B 9012)

30 Much of the Crean family information was graciously provided by Richard Crean of Toronto, in consultation with Irish relatives.

31 *Saskatchewan Archives Board* (*SAB*), Crean papers, F.J.P. Crean to R.E. Young, 24 April 1908.

the Manitoba legislature. Crean likely knew Robert Young from his surveyor days in western Canada, and appears to have been recruited with the Saskatchewan expedition in mind; he no sooner joined the Railway and Swamp Lands Branch on 4 February 1908 than he was asked for his views on heading an exploratory party to the region.

Crean's assessment of the task at hand was articulate and reasoned. After reviewing what was known about the district and the need for a survey, he recommended that everything about the region be noted and reported upon; he also believed that the camera was indispensable to this exercise, especially if the findings were to be published. "To the lay mind," he observed, "illustration is a most convincing if not conclusive argument."[32] Young undoubtedly appreciated this argument, for his sole hobby, apart from northern propagandist, was photography. Drawing upon his northern field experience, Crean also advised that the survey be conducted mainly by canoe, the equipment be kept to a necessary minimum, and the provisions be supplemented by supplies at the local HBC posts. As for the size of the party, he insisted that it be kept small—no more than two assistants—and that the men's willingness to take orders was more important than experience. His worries about discipline, however, were unfounded. At six foot, the heavy-set, moustached Crean was an intimidating figure, all the more so since he sported a powder burn around his right eye and on his throat; he was reportedly referred to by Indians as "Big One Eye." It was also readily apparent from his concluding remarks that Crean understood the larger significance of the expedition and what it meant to Young. "It is practically certain," he commented, "that this tract of country is convertible into a fertile land and apparently all that is required is to give the Canadian farmer an idea of its possibilities."[33] Young had found his disciple.

Crean wanted to leave for the field as early as possible. It was not until early August, however, that he was finally told by Young that he could proceed to Prince Albert. The reason for the delay is not clear; perhaps the general election campaign made it difficult to secure government approval. According to Young's instructions, Crean was to explore all the lands lying south of the Churchill (English) River and between Green Lake/Beaver River on the west and Cumberland House/Stanley Mission on the east; the exact route to be followed was left to Crean's discretion on the under-

32 *Ibid.*

33 *Ibid.*

standing that he was to concentrate on areas away from well-travelled routes. He was assigned one assistant, William Ronald Caldwell, a twenty-one-year-old clerk from the same branch. Together, with whatever temporary local help was hired in the field, they were to examine and evaluate all aspects of the region and its resources. To facilitate this work they were given two cameras and told to take photographs at every opportunity, especially images of any crops in the district. Crean was also instructed to keep a daily diary, which unfortunately has not been located. When winter set in, Crean was expected to cruise any major timber stands in the area before returning to Ottawa by 1 January 1909.[34] In recognition of the importance of the expedition and possibly as an added incentive, Young arranged for a substantial increase in the monthly salaries of Crean and his assistant just prior to their departure.[35]

Crean spent the middle part of August in Prince Albert, getting together the expedition outfit, as well as consulting with various local officials. Most people scoffed at the idea of a large tract of farming land to the north and suggested that he would have better luck finding precious metals. He was also told by Robert Hall, district manager of the Hudson's Bay Company store, that his plan to hire only two packers was foolhardy, especially since he would be working inland, and that he should take on at least four experienced men and expect to change them at regular intervals to ensure the best possible local knowledge of the country. Crean took the advice and wired Young for permission to enlarge the expedition party; he also asked for an additional $1,000, in the form of a credit with the Hudson's Bay Company, to supplement the original $1,500 budget for the expedition. While waiting for approval of these matters, Crean hired a team and buggy and investigated the crude trail that led north from Prince Albert to Montreal Lake. Finding the road almost impassable in places, he decided that the expedition would instead go northwest to Beaver Lake and Île-à-la-Crosse and then work its way back east to Lac La Ronge.[36]

The Crean expedition officially got underway on 20 August 1908, when it headed northwestward from Prince Albert along the Green Lake Trail. It was entering a

---

34 *Ibid.*, R.E. Young to F.J.P. Crean, 6 August 1908.

35 *NAC*, Government Archives Division, Department of the Interior, RG 15, v. 1030, f. 1645827, W.W. Cory to Beddoe, 17 August 1908.

36 *Ibid.*, 1 August 1908. The date of this letter is clearly wrong since Crean did not leave Ottawa for Prince Albert until after 6 August.

region already rich in history. In the late eighteenth century, the Churchill and Saskatchewan river systems had served as a commercial battleground, as the Hudson's Bay Company and the Montreal-based Northwest Company struggled for economic supremacy in the Canadian fur trade. At first, the Indians of the region, mostly Woodland Cree and Chipewyan, were able to take advantage of their position to dominate the early trade in the region, but as the number of posts along the river corridors proliferated and competition became ruthless, at times violent, their role as middlemen was effectively undermined. The bitter rivalry finally came to an end with the amalgamation of the two companies under the HBC name in 1821; economy, consolidation, and rationalization became the order of the day. In the Saskatchewan/Churchill country, this streamlining eliminated duplicate posts and personnel, and the region generally became more important as the major trade route into the fur-rich Athabasca country. All traffic to Fort Chipewyan and the Mackenzie Valley was normally routed along the Churchill River via Île-à-la-Crosse and Portage La Loche (Methye Portage). The Cree, Chipewyan, and mixed-blood populations, in the meantime, were encouraged by the remaining centres not only to continue to trap furs but also to harvest country provisions for the distant posts and canoe brigades going in and out of the far northwest.

Northcentral Saskatchewan had also been the scene of an intense mid-nineteenth-century rivalry for Aboriginal souls. In 1846, James Settee, an ordained Native, established a Church of England mission at the west end of Lac La Ronge; it was later relocated to Stanley Mission on the Churchill River. The building of Holy Trinity Church in 1856—the oldest existing structure in Saskatchewan—initially led to the Anglican domination of the northeastern side of the province. The northwestern side, on the other hand, was essentially a Roman Catholic stronghold. Oblates Louis Laflèche and Alexandre Taché established a permanent mission (St. Jean Baptiste) at Île-à-la-Crosse in 1846 and within two decades, mass was being celebrated throughout the area from Portage La Loche to Reindeer Lake. Over the next half century, the Île-à-la-Crosse mission became known as the "Bethlehem of the North," especially with the arrival of the Grey Nuns of Montreal in 1860 and the building of a convent, orphanage, and school. It was also called the "nursery of bishops" because four future bishops (Laflèche, Taché, Grandin, and Faraud) did missionary work there.[37]

Like many explorers before him, Crean would depend on the goodwill and assistance of the local fur trade posts and missions; their knowledge of the land and its people would prove invaluable. The days of the Old Northwest, however, were drawing to a close in the late nineteenth century. The transportation system through northcentral Saskatchewan was abandoned in the mid-1880s in favour of a new route north from Edmonton, while a dramatic rise in the price of furs brought scores of independent traders into the region.[38] The greatest challenge to the venerable HBC came from the Paris-based Revillon Frères, which opened a district office in Prince Albert in 1901 and established a number of competing posts throughout northern Saskatchewan over the following three years. In the meantime, the Canadian government appeared to be taking its Indian responsibilities more seriously when it embarked on a second round of treaty-making in 1889. On a bitterly cold day in February of that year, the Montreal Lake and Lac La Ronge bands signed an adhesion to Treaty 6 (originally negotiated 1876). Ten years later, the Chipewyan of northwestern Saskatchewan were included as part of Treaty 8, which covered present-day northern Alberta, northeastern British Columbia, and a wedge of the Northwest Territories south of Great Slave Lake. Then in 1906, one year after Saskatchewan became a province and two years before the Crean expedition took to the field, the rest of northern Saskatchewan was surrendered in Treaty 10 (except for the Cumberland House area, which was part of Treaty 5 (1875)). It is debatable whether Ottawa was simply trying to ensure that the Indians would be able to continue to pursue their harvesting lifestyle or was preparing the way for future settlement and development of the region.[39]

The arrival of the Qu'Appelle, Long Lake, and Saskatchewan Railway in Prince Albert in 1891 was also a symbol of the changing times. The railroad provided a much-needed outlet for the region's agricultural produce and in so doing, encouraged hundreds of new settlers to take up farm land. It also rekindled some of the earlier expectations for Prince Albert, as local entrepreneurs now moved to exploit the resources of the nearby boreal forest to supply southern markets. By 1900, several hundred thousand pounds of fish were being taken an-

37 Fr. G. Carrière, "The Early Efforts of the Oblate Missionaries in Western Canada," *Prairie Forum*, 1979, v. 4, n. 1, 5–9; P. Duchaussois, *Mid Ice and Snow: The Apostles of the North-West* (London 1923), 91.

38 H.J. Moberly (*When Fur was King* (Toronto 1929), 174–6) recalled that the Stanley Mission/Île-à-la-Crosse area was overrun with free traders at the end of the century.

39 See K.S. Coates and W.R. Morrison, *Treaty Ten (1906)* (Ottawa 1986).

nually from northern lakes and exported to the United States. Lumbering also experienced phenomenal growth during these years. Prince Albert's sawmills supplied an increasing share of the western Canadian lumber market—16 percent in 1904—as their winter cutting operations steadily pushed the timber frontier northward through the virgin white spruce stands. In fact, although Crean had once been one of their own, local lumbermen were probably somewhat wary of his assignment and worried that his findings might possibly lead to limits or controls on their activities; as far as they were concerned, the region was best suited for lumbering, not settlement.[40]

Crean's early findings suggested a completely different picture. As the expedition worked its way northwestward along the 150-mile Green Lake Trail,[41] it passed field after field of grain in the recently opened Shellbrook and Mont Nebo districts; Crean was, in fact, so impressed that he sent a wheat sample to Young. Beyond Big River, the country became more rolling, broken, and treed, but the presence of abundant hay in the beaver meadows convinced Crean that it would be ideally suited for ranching. The expedition reached the freight depot at the south end of Green Lake in five days—the last two through steady rain—and then pushed on by canoe to the HBC post at the opposite end of the lake. At the Green Lake settlement, Crean examined the cultivated fields and gardens of the local Métis farmers and Roman Catholic mission, and travelled on horseback as far east as Pelican (Lavallée) Lake in the northwestern corner of present-day Prince Albert National Park. He also interviewed several local residents about their farming experiences in the country, and although encouraged by their stories of repeated success, he was somewhat frustrated by the fact that no one bothered to keep a record of seeding and harvesting dates or crop yields. Crean sent Young a brief report of his findings on 1 September, in which he claimed that the land around Green Lake was "first class"[42]

40 B. Waiser, *Saskatchewan's Playground: A History of Prince Albert National Park* (Saskatoon 1989), 11–14.

41 The Green Lake Trail had been opened by the Hudson's Bay Company in the mid-1870s to improve access to its northern posts along the Churchill transportation corridor and farther west. It had been surveyed by chain and transit by J. Burgeois of the Department of the Interior in 1888.

42 Crean would later report (*Northland Exploration*, 30) that a "representative" quarter section in the region could be broken down as follows: good land prairie—8 acres; good land bush, not hay—80 acres; hay land (not requiring drainage)—15 acres; hay land in need of drainage—20 acres; muskeg (probably possible to drain)—10 acres; muskeg (difficult or impossible to drain)—10 acres; stony land—2 acres;

and that the country to the west was reportedly even better. He also lamented the time allowed for his survey—"I can only take a bird's eye view . . . of what I am firmly convinced is a great land"[43]—and sought Young's permission to winter over and continue his assessment by dog team.

From Green Lake, the Crean party paddled north down the Beaver River towards Île-à-la-Crosse, using makeshift sails whenever possible to speed their progress. At the confluence of the La Plonge and Beaver rivers, they briefly inspected the new Roman Catholic mission,[44] where they found a small sawmill and one of its recent beneficiaries, a large three-storey school nearing completion. What was even more remarkable was a field of oats that reached Crean's shoulder and a fine garden with an astonishing variety of vegetables; a small patch of wheat also appeared promising but had been damaged by frost.

On 5 September, the Crean expedition reached Île-à-la-Crosse. Although the post and mission were renowned in fur trade and church circles, the site seemed poorly suited for agriculture because of the low, swampy character of the land; in fact, many of the supplies for the HBC post and its Revillon counterpart were now freighted in from Prince Albert. Crean was much more enthusiastic about the potential of the land to the northwest around Buffalo (now Peter Pond) and La Loche lakes, which he visited in mid-September. At a small Chipewyan settlement at the mouth of the Buffalo River (present-day Dillon), he photographed a large vegetable plot and small patches of oats and barley. He also spent several days at La Loche, the former HBC trans-shipment centre to the Mackenzie Basin, where he learned that the region had once supported considerable mixed farming but that much of this activity had come to an end when the route was abandoned.[45] He was also told by the local Roman Catholic mission that its fields were necessarily small because of the lack of sufficient seed. While at La Loche, Crean walked across the famous Methye Portage and concluded that the Clearwater River valley was comparable to the North Saskatchewan and would one day be a magnificent cattle range.

Crean also explored westward from Île-à-la-Crosse. At the Canoe Lake mis-

---

water (small ponds)—15 acres; for a total of 160 acres.

43 *SAB*, Crean papers, F.J.P. Crean to R.E. Young, 1 September 1908.

44 In 1912, the La Plonge mission was renamed Beauval to avoid confusion with La Ronge.

45 During the Klondike Gold Rush, the route was temporarily re-opened by the HBC to assist stampeders trying to reach the Yukon that way.

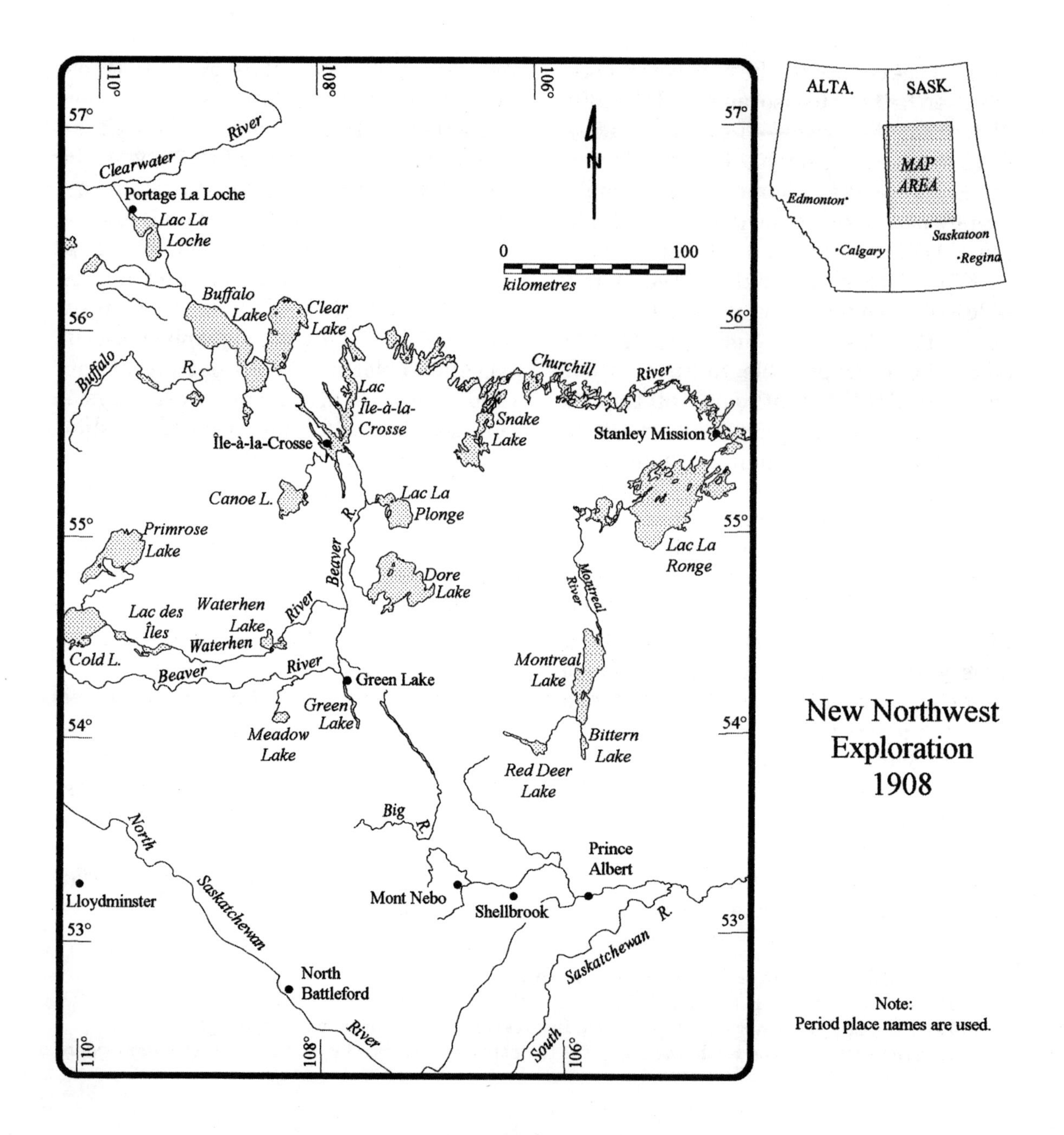

New Northwest Exploration 1908

Note:
Period place names are used.

sion (present-day Canoe Narrows) in early October, he found some good gardens but decided that little of the region was fit for farming because of the prevalence of muskeg. He reached a similar conclusion about the land between Canoe and Buffalo lakes, which he slogged through only with great effort. In fact, Crean provided a more sober assessment of the country's potential in a hurried letter to Young in mid-October. "I have seen nothing startlingly good," he confided, "still I am convinced that it is by no means waste land."[46]

Crean spent the better part of September and October of 1908 at Île-à-la-Crosse, which more or less served as the unofficial expedition headquarters. He probably would have travelled more extensively in the region but found it difficult to secure guides who were willing to go inland away from the travelled routes; he and his assistant, Caldwell, consequently made little exploratory trips on their own. Even then, overland travel proved arduous, and Crean anxiously looked forward to freeze-up when he could travel faster and straighter by dog team. During his time at Île-à-la-Crosse, Crean took some simple scientific observations and kept a daily record of the weather, including the nightly minimum temperature.[47] He also took photographs of the local inhabitants, who proved willing subjects even when the camera was empty, and developed all of his exposed film.[48]

Although Crean expected to leave Île-à-la-Crosse in mid-October, his departure was continually delayed because of the lack of snow. It was consequently not until 20 November that he struck out by dog team for Snake Lake (now Pinehouse) on the Churchill River system and rendezvoused with Caldwell, who had been sent ahead with the provisions via the conventional water route.[49] The men then continued southeastward across country until they hit the Montreal River, where they swept north around Lac La Ronge to Stanley Mission. At the time of the visit, there was considerable speculation about potential mineral deposits in the region. Crean doubted, however, that the mines would ever amount to anything and chose not to secure any samples, since William McInnes of the Geological Survey had just

46 *SAB*, Crean papers, F.J.P. Crean to R.E. Young, 13 October 1908.

47 After Crean left Île-à-la-Crosse, these recordings were continued by A.H. Peirce of the HBC and were reproduced in the final expedition report.

48 *SAB*, Crean papers, F.J.P. Crean to R.E. Young, 13 October 1908.

49 Crean had "a shack of sorts," which he likely used for trapping, where the Sandy River entered Snake Lake. *Ibid.*

been through the region. Instead, he placed more faith in the productivity of the soil, especially after he read in the HBC post diary that wheat at one time had been grown successfully enough to merit construction of a mill at the mission.

From Stanley, Crean and his assistant continued southward and after being forced to sit out a blizzard, reached the HBC post at the south end of Montreal Lake in early December. Here, Crean intercepted an official letter that had been addressed to him at Lac La Ronge in response to his earlier request from Green Lake that he be allowed to winter over to continue his survey. The news from Ottawa was brief and to the point: Because funds for the expedition were fixed at $2,000, the only thing that Young could do was extend the date of Crean's return to Ottawa to the end of February 1909.[50] Crean immediately replied that he had "covered the country allotted to me fairly well" and that he would make one more run through the heart of the survey area before coming out. He also included a somewhat contradictory assessment of the Lac La Ronge district, which only confused the question of the region's agricultural potential. "My opinion of this country so far," he advised Young, "is that through much human energy it would produce a great quantity of grain and vegetables still it is not a farming country."[51]

The expedition headed directly west from Montreal Lake to Green Lake on a course that took it around the three largest lakes (Crean, Kingsmere, and Waskesiu) in present-day Prince Albert National Park. It was while running along the trapping trails or "pitching tracks" in the region that the first serious accident befell the survey, when Caldwell—in Crean's words, "the toughest of the tough"[52]—severely twisted his knee and had to be sent out to Prince Albert. Crean continued alone, and although the land around the lakes was deep in the grip of winter, he would later report, "I fancy this country might profitably be surveyed and opened for settlement."[53] From Green Lake, he travelled north to Île-à-la-Crosse, where he disposed of the expedition's canoes and tents and closed his HBC account. Crean eventually returned to Prince Albert on 6 January and wired for an additional $250 to cover his overdrawn account and more importantly, to request a leave of absence—he was getting married thirteen days

50 *Ibid.*, J.B. Challies to F.J.P. Crean, 24 September 1908.

51 *Ibid.*, F.J.P. Crean to R.E. Young, 7 December 1908.

52 *Ibid.*

53 Frank J.P. Crean, *New Northwest Exploration* (Ottawa 1910), 43.

later to Josephine Mary Gigot, the eldest daughter of Edward Gigot, the HBC store manager in Nelson, British Columbia. By sheer coincidence, on the streets of Prince Albert, Crean met a former school chum, Captain F.T. O'Meagher, who agreed to serve as his best man.[54]

Crean submitted his formal expedition report to Robert Young on 10 April 1909. It was divided into two parts: a general summary of the region's resources and a more detailed section-by-section description of the survey area. Crean couched his findings in such a way that greater emphasis was placed on the land's potential than on what he actually saw. After noting that his exploration was "necessarily hurried" and that the whole tract was "not exactly fitted for agricultural settlement throughout in its present state,"[55] he explained that the climate was ideal for raising any kind of cereal and that wheat had been grown successfully wherever it had been tried. He also suggested that almost one-quarter of the survey region's 22 million acres was suitable for cultivation *once* the land had been surveyed and made accessible.[56] And much of what was not good for mixed farming could be used for forestry or mining. Crean also deplored the amount of valuable timber that was lost each year to careless fires, and intimated as well that the local Indian and mixed-blood population, by relying largely on the diverse game and fish resources of the region, were not taking advantage of the potential of the soil.

Young forwarded Crean's report to Frank Oliver on 1 May and secured approval from the minister of the interior to have it reproduced for public distribution. Entitled *Northland Exploration* (Ottawa 1909), the pamphlet carried an endorsement by Professor John Macoun of the Geological Survey, whose enthusiasm for the northland knew no limits. It was also lavishly illustrated with many of the photographs that Crean had taken during his travels. The splendid pictures of the northern gardens and fields—even though they were but a few acres in a vast

54 Nelson *Daily News*, 20 January 1909; while at La Loche, Crean had arranged for the manager's wife, Mrs. Groat, to do some special "fancy work" for his new bride. *Hudson's Bay Company Archives*, B.85/b/24, A. McKay to R.H. Hall, 18 February 1909.

55 Crean, *New Northwest Exploration*, 16.

56 Crean's Irish background may have figured in his enthusiastic assessment of the region, especially since his father worked for the Irish Land Commission *after* the Irish land reforms. The commission was largely concerned with assessing the agricultural potential of particular pieces of farm land and the value of improvements made by tenants. By the standards of the Irish Land Commission, Crean's assessment was probably more reasonable than it seemed when assessed in a Canadian context.

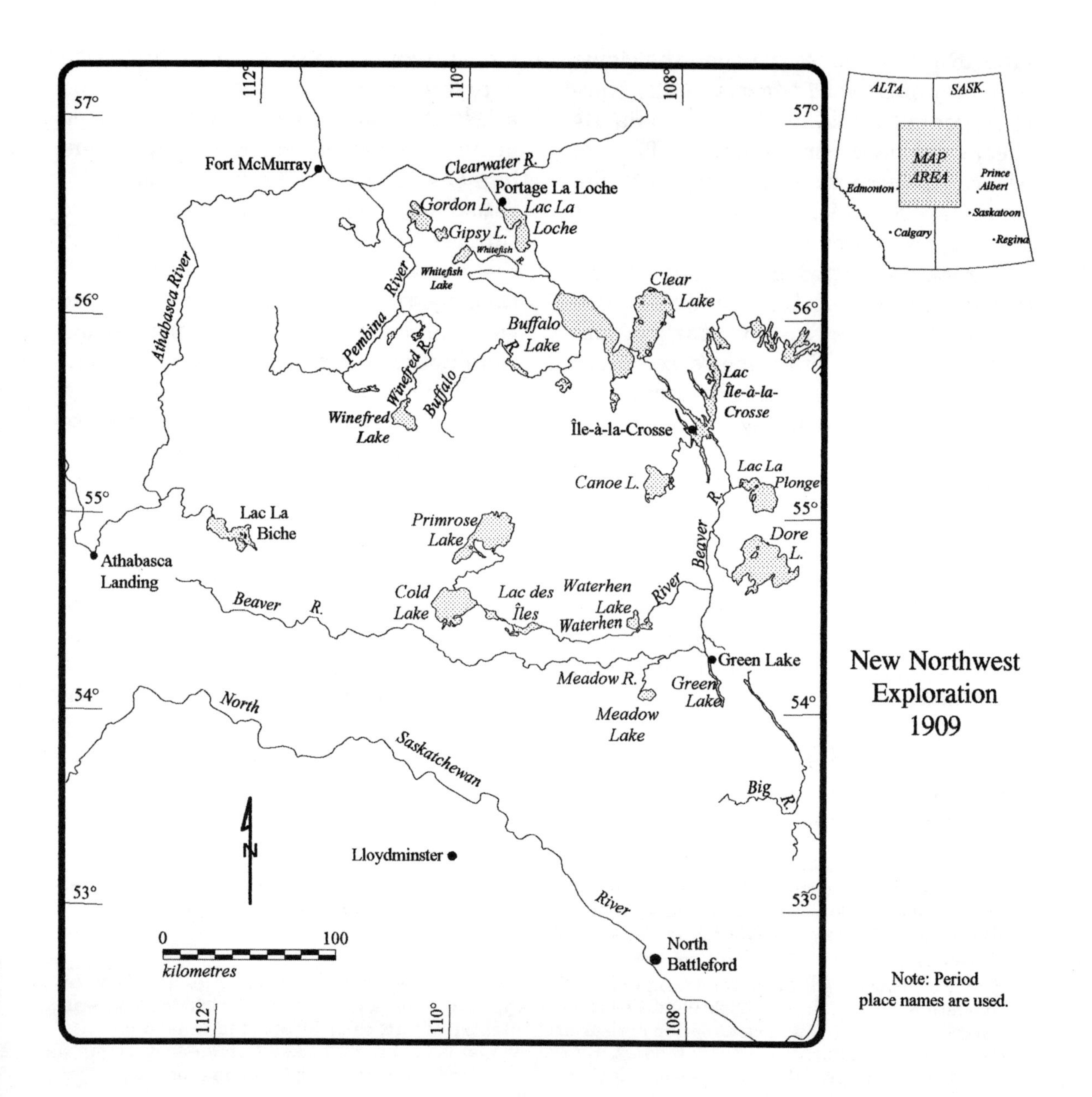

New Northwest Exploration 1909

Note: Period place names are used.

land—were eloquent testimony to the region's promise.

As in the case of *Canada's Fertile Northland*, the Crean pamphlet was freely available upon request. The Department of the Interior also sent copies to newspapers, which raved about the expedition and its findings. Not a single article raised any questions or doubts; the only complaint was that the government should be doing more to advertise the region.[57] Young's promotional efforts were also helped by Agnes Deans Cameron, an American school teacher who had travelled through northern Canada during the summer of 1908 and was now wowing audiences across North America and Great Britain with her lectures on "the new north."[58] Young had also been busy behind the scenes trying to convince government surveyors and scientists to produce popular versions of their official reports.[59] There was even talk of sending the visiting members of the British Association for the Advancement of Science on a northern excursion following their annual meeting in Winnipeg.

While all the fuss was being made about Crean's findings, he was already back in the field. On 18 May 1909, less than three weeks after his report had been submitted to the minister, he had been instructed by Young to leave for Prince Albert as soon as possible and continue his exploration into northeastern Alberta as far north as the Clearwater River and as far west as the Athabasca River. One month later less a day, the second Crean expedition set off along the Green Lake Trail. It was a considerably larger party than the year before; Crean was accompanied not only by a new department assistant, twenty-four-year-old Alfred Martin Beale, but had also been allowed to hire five additional men in Prince Albert.[60] Its mandate, however, remained the same: Observe everything and in so doing, confirm the great expectations for the region.

The expedition's headquarters for the first two months was the Hudson's Bay

57 Victoria *Times*, 6 August 1909; Edmonton *Journal*, 7 August 1909; Manitoba *Free Press*, 7 August 1909; St. John *Globe*, 7 August 1909; Ottawa *Free Press*, 7 August 1909; Vancouver *World*, 7 August 1909; Edmonton *Bulletin*, 10 August 1909; Regina *Leader*, 11 August 1909; Prince Albert *Herald*, 13 August 1909; Toronto *Globe*, 14 August 1909; London *Advertiser*, 19 August 1909.

58 See Agnes Deans Cameron, *The New North: An Account of a Woman's 1908 Journey through Canada to the Arctic* (1909. Reprint. Saskatoon 1986), edited and with an introduction by David Richeson.

59 Young carried on an extensive correspondence with J.B. Tyrrell regarding a pamphlet on the barren lands, tentatively titled, "Treeless Plains of Northern Canada." *TFRBL,* Tyrrell papers, Box 92, J.B. Tyrrell to R.E. Young, 4 January 1909.

60 T.G. Street, F.T. O'Meagher, J. Brown, H. Rosson, G. Martin (cook).

Company post at the north end of Green Lake.[61] From here, Crean explored west along the Waterhen River and the series of lakes that make up present-day Meadow Lake Provincial Park. The land near the junction of the Beaver and Waterhen rivers was found to be extremely swampy and virtually impassable to the northwest. But as the party worked its way westward to Lac des Îles, Crean was greatly impressed by the fine hay lands and scattered stands of spruce along the route. He also concluded from the abundance of peavine and vetch that the land must be good—even though no crops were being grown in the district—and that this good land probably extended beyond the fourth meridian into Alberta.

Since a survey team was already working in the Cold Lake region laying out the Saskatchewan–Alberta boundary, the Crean party portaged south from Lac des Îles to the east-west arm of the Beaver River, proceeded east to the Meadow River and then south to Meadow Lake. As they paddled into the settlement on 1 August, it must have seemed like an oasis; not only was the land around the lake mostly prairie—hence its name—but it supported the beginnings of a thriving agricultural community. There were prolific gardens, small herds of cattle and horses fattening on the local grass, and an impressive field of Banner oats at the farm instructor's house on the nearby Indian reserve. Crean's investigation of the surrounding region by horseback served only to confirm his initial impressions, and he would later declare that the district contained "some of the very finest farm land in Canada."[62]

Once Crean had completed his assessment of the land west of Green Lake, the expedition continued north to Île-à-la-Crosse and then on to La Loche, its new base of operations. Over the next few months until freeze-up, Crean conducted most of his field work on the Alberta side of the border, inspecting the land south of the Clearwater River as far west as its junction with the Athabasca. As during his 1908 survey, he was also under explicit instructions to avoid established water routes or trails in favour of travelling through lesser known tracts in order to obtain a more complete picture of the resources in the survey district. In a sense, Crean and his party could be regarded as "original observers of the land"[63] in that they were evaluating the potential use of

61 The exact route followed by the expedition is uncertain (unlike 1908, there is no surviving correspondence) and has been pieced together from Crean's subsequent report.

62 Crean, *New Northwest Exploration*, 78.

63 D. Owram, *Promise of Eden: The Canadian Expansionist Movement and the Idea of the West* (Toronto

the subarctic under completely different circumstances and objectives than attended fur trade and missionary activity in the region. And it would be difficult if not impossible to be entirely objective, given the new expectations for the region.

In late August, Crean and his party set off for nearby Whitefish (now Garson) Lake by canoe instead of using the more direct, summer cart trail. It was a decision they quickly regretted since they had to drag their boats for several miles along the shallow, snag-ridden Whitefish (Garson) River. Their efforts, in the end, were also poorly rewarded; except for some small plots of root vegetables at the settlement, the land around the lake was generally swampy and in need of drainage. These wet conditions continued to plague their travel as they worked their way north via Gipsy Lake over the height of land to the Clearwater River and the arctic watershed. By the time they finally escaped the dreary country to the south and stood on the edge of the Clearwater valley, Crean felt a mixture of relief and exhilaration.

During his 1908 survey, Crean had briefly crossed over to the Clearwater and come away highly enthused about the region's future. Now, as the expedition travelled down the river towards the Athabasca, he became more than ever "convinced that to this valley nature has been extremely bountiful."[64] There were large hay meadows on the valley benches, untouched stands of timber on the slopes, and power waiting to be harnessed in the river basin. If he had any lingering doubts that the valley would not support a large agricultural population, they were easily put to rest by the field of wheat that he found at Fort McMurray. All the region needed, according to Crean, was a railway connection from Edmonton.

From Fort McMurray, Crean and his men returned inland up the Pembina (now Christina) River, a tributary of the Clearwater. Around Cowper Lake, they found a wide band of open prairie land that seemed ready for immediate occupation. That the tract had promise was confirmed by a local band of Chipewyan Indians who sowed gardens on the prairie in the spring and did not return again until the fall to harvest the produce. Such practices amazed Crean, who not only marvelled at how the neglected gardens throughout the northland, including those at fur trade posts, produced prolific crops, but wondered what the yields would be like if the plots were ever properly tended. At the

1980), 65. See also J. Warkentin, "Steppe, Desert and Empire," in A.W. Rasporich and H.C. Klassen, eds., *Prairie Perspectives 2* (Toronto 1973), 128.

64 Crean, *New Northwest Exploration*, 83.

same time, he knew that until there was a larger population in the area and better access, there was no incentive to plant crops, let alone experiment.

As the expedition pushed south towards Winefred Lake, it became bogged down in the swampiest hay land that it had encountered on the Athabasca watershed. But beyond this wet tract, Crean generally found the country from the interprovincial border towards Lac la Biche to be ideally suited for the pioneer. Although slightly rolling and more broken, it was amply endowed with hay, wood, water, and fish. In fact, in Crean's mind, it was comparable to the Edmonton district, and he saw no reason why it should not be opened up to mixed farming.

By the early fall, the expedition had returned to La Loche. Crean released all the Prince Albert men except T.G. Street, who agreed to stay on as his winter travelling companion. His departmental assistant, Arthur Beale, also returned to Ottawa at this time, probably to work up some of the field data, as well as make a preliminary report to Robert Young. While waiting for freeze-up, Crean visited some of the larger lakes around La Loche and witnessed the annual fall fisheries; on one day alone, twelve thousand fish were hung at the north end of Buffalo (Peter Pond) Lake.[65] He also likely cruised any major timber stands in the area. It is not known when Crean and Street left La Loche by dog team, but they first travelled west to Fort McMurray and then south to Edmonton, arriving there on 11 December 1909. By the time Crean returned to Ottawa, he had spent seven months in the field.

Crean's official report of his 1909 activities, submitted in early May 1910, was even more favourable than the previous year's findings. He estimated that almost half of the 21 million acres within the survey region was suitable for settlement, and that much of the remaining area could be profitably drained. He also portrayed the region, with its wide range of natural resources, as being better suited for settlers of limited means than the open prairie to the south. "Although the North may never seriously compete with the more southerly latitudes in the wheat market," Crean observed, "still, by judicious mixed farming, it will eventually be equally productive and support a dense, thriving population."[66] He also emphasized, as he had done a year earlier, that the single most important obstacle to large-scale colonization was access to the region—and not the climate. He was quite insistent that any agricultural activity in

65 Crean, *New Northwest Exploration*, 73.

66 Crean, *New Northwest Exploration*, 69.

the survey district, however limited, had always proven successful.

These findings greatly pleased Robert Young and he evidently decided to send Crean and Beale north again later that summer, this time to assess the region beyond the Clearwater River and east of the Athabasca and Slave rivers.[67] They would be following in the wake of the minister of the interior, Frank Oliver, who was making—most likely at Young's urging—his own personal tour of the region on his way to the Yukon. As part of his ongoing promotion program, Young also arranged to have the two Crean reports published together—complete with photographs and maps—under the title *New Northwest Exploration*. In addition, he commissioned Ernest Chambers to prepare an edited compilation of all the available information on the resources of the Northwest. Young's star, in the meantime, was in the ascendant. In February 1909 he served as secretary to the Canadian delegation at the North American Conservation Conference in Washington, D.C. The following spring he was named chief geographer for the dominion, in addition to his regular duties. He derived his greatest satisfaction, however, from his work on behalf of Canada's northland and took tremendous pride in the fact that his great faith in the region's future had been confirmed by the Crean expeditions.

This vision of a prosperous, settled northland was not to be realized. In 1911 a new Conservative government assumed office, and the transfer of power was no sooner completed than the boom years of the preceding decade-and-a-half gave way to a recession that ate away at the economy until after the outbreak of the Great War. Western Canada was particularly hard hit. Grain prices plummeted, the construction industry, and with it the real estate market, collapsed, and the two new, uncompleted transcontinental railways sustained mortal blows from which they never recovered. There was even drought in some southern districts. It was not a time, then, to be contemplating the opening up of the northland when the

67 Any material on this third expedition is extremely vague and circumstantial. There is no reference to the expedition in the annual reports of the Department of the Interior—nor for that matter, any mention of the 1908 and 1909 expeditions. According to surviving Interior records, Crean bought a scow from Father Gouy at the Roman Catholic mission at Smith Landing in September 1910 (*NAC*, Department of the Interior, RG 15, v. 1030, f. 1645827, A.J. Arthur to F.C.C. Lynch, 5 December 1913). In the Crean papers at the Saskatchewan Archives Board, Beale mentions in a brief autobiographical note that he returned to Ottawa in early 1911 after *two* seasons' field work in northern Saskatchewan and Alberta. Finally, in a 1914 map of northeastern Alberta, present-day Leland Lakes (just east of Smith Landing) are named Beale Lake.

prairies were experiencing such trouble; there was no longer any urgency. This was clearly evident by the way in which Ernest Chambers's latest book on the region, *The Unexploited West*,[68] was handled. Ready by May 1912, it was not published by the government for another two years.

It was also found that Crean's conclusions could not stand up to scrutiny. Surveyors sent into the region north of Prince Albert in 1909 and 1910 to lay out major base lines, such as the third meridian, came across little land of agricultural value.[69] Faced with these contradictory findings, the Department of the Interior in 1912 commissioned C.H. Morse of the University of Toronto School of Forestry to conduct another reconnaissance of the district lying between the third meridian on the east and the Big River branch line of the Canadian Northern Railway on the west. Morse went about his assignment in business-like fashion, carefully noting the region's drainage, topography, soil composition, climate, and vegetation, including the amount and type of merchantable timber. At season's end, he recommended that the land east of the Sturgeon River be set aside as a federal forest reserve. He found the tract to be characterized by rolling, heavily timbered, gravelly or sandy soils, and therefore poorly suited for agriculture, as evidenced by the relative lack of settlement.[70] These findings were corroborated two years later when Assistant Forester L. Stevenson of Prince Albert undertook a soil survey of the region. "It would be very unfortunate," he warned, "if this area were opened for settlement, as nothing but hardship and poverty followed by starvation awaits those people so constituted as to desire to settle in lands that have a better use in forestry."[71]

Such assessments were damaging in themselves, but what made them even more damning was that they went largely unchallenged.[72] Little was said, however,

68 E.J. Chambers, *The Unexploited West: A compilation of all available information as to the resources of Northern and Northwestern Canada* (Ottawa 1914).

69 See, for example, SAB, RG 15, Department of the Interior, Surveyors' Reports, 1457, A. Saint Cyr to E. Deville, 9 February 1910.

70 *NAC*, Government Archives Division, RG 39, Forestry, v. 451, f. 40590, R.H. Campbell to W.W. Cory, 2 April 1913.

71 *Ibid.*, L. Stevenson to G.A. Gutches, 10 September 1914.

72 In 1912 Captain James King of Prince Albert published a pamphlet, popularly known as *The Settler's Guide*, which championed the settlement potential of northern Saskatchewan while downplaying any pessimistic assessments. J. King, *A Guide to the Great Possibilities of Northern Settlement in Saskatchewan* (Prince Albert 1912).

because Robert Young was gone; on the morning of 24 October 1911, he had been found in a sitting position in his bed, dead from a heart attack, with *The Life of Gladstone* still open in his hands.[73] Young's death robbed the northland of its greatest promoter, and the region would never again enjoy the same government attention and national fascination. Ernest Chambers said as much in his introductory remarks to *The Unexploited West,* when he referred to Young as "the most enthusiastic believer in the northland who ever lived."[74] This legacy was belatedly recognized in 1917, when the Railway and Swamp Lands Branch of the Department of the Interior, in keeping with its actual activities, was renamed the Natural Resources Intelligence Bureau.[75]

Frank Crean, for his part, resigned his government position in February 1913 and briefly served as an agent for the Canada Life Assurance Company in Calgary before securing work with his brother-in-law in High River. In January 1915 he enlisted at Calgary in the Mounted Rifles (12th Regiment) for overseas service and then abruptly requested his discharge a month later. Crean was then commissioned as a lieutenant in the active militia (12th Manitoba Dragoons) and spent the remainder of the year at the Brandon internment centre for enemy aliens before being removed from duty because of drunkenness. He then re-enlisted for overseas service and joined the Calgary Number 2 Tunnelling Company in France in March 1916. His drinking, however, once again interfered with the performance of his duties—his commanding officers politely described him as incompetent—and he was sent back to Canada at public expense in October. Upon debarkation, he continued his antics by failing to report and was not heard from for several months.[76]

After the war, Crean returned briefly to Ottawa, presumably to try to find employment with the Department of the Interior. He then re-located to Toronto and apparently lived there—except for a brief period in Belle River, Ontario, in the late 1920s—drifting from job to job for the remainder of his life. At the time of his death on 18 December 1932, the fifty-seven-year-old Crean was again out of work and was buried by the Last Post Fund in Mount Hope Cemetery. His widow, Josephine, later moved to Victoria to be near family, and was killed accidentally in 1963 when she

73 Ottawa *Citizen*, 25 October 1911.

74 Chambers, *The Unexploited West*, viii.

75 See I. Spry and B. McCardle, *Records of the Department of the Interior* (forthcoming, Regina 1993).

76 Information about Crean's war record was secured from his military personnel file.

was struck by a car on her way home from mass.[77] The couple had no children.

As for the New Northwest, it became the land of a second chance in the interwar period. As part of a reconstruction program, over two hundred new townships along the southern edge of the boreal forest were opened up to homesteaders, especially veterans, after the Great War. And during the Depression, many settlers abandoned their land in the drought-stricken south and headed north to start over again—what one author has described as "the great trek north."[78] Farming in the region, however, proved to be a difficult, extremely risky activity and could never support the kind of population and development envisaged by Young and endorsed by Crean.[79] In fact, the very idea of a fertile northland appears in retrospect to have been a ridiculous, even reckless, proposition. At the same time, it is important to realize that the notion was representative of a period in Canadian history when there was an unparalleled belief in the country and its destiny. And this destiny came to include the more remote parts of the Northwest; it had always been assumed that the region had hidden potential and would some day prove valuable. In the early twentieth century, as western Canada was being populated by a seemingly endless stream of immigrant farmers, settlement of the land north of the North Saskatchewan was seen as the next stage in the gradual rolling back of Canada's agricultural frontiers. The Old Northwest was to be replaced by a newer, improved version, which, according to Young, was essential to the continued growth of the country.

The photographs in this collection should be viewed in this context—the Old and the New Northwest. When Frank Crean travelled through northwestern Saskatchewan and northeastern Alberta in the early 1900s, he captured with his camera the final days of the Old Northwest and its unique way of life. The Indian and mixed-blood settlements, fur trade posts, and missions had developed in relative isolation during the nineteenth cen-

77 Victoria *Daily Times*, 15 March 1963.

78 See D.P. Fitzgerald, "Pioneer Settlement in Northern Saskatchewan," unpublished Ph.D. thesis (geography), University of Minnesota, 1966.

79 The estimated 35,000 to 45,000 settlers who sought refuge in the forest fringe of northern Saskatchewan between 1930 and 1936 eked out a miserable existence on submarginal land. It is noteworthy that most of this settlement occurred *south* of the region examined by Crean. See J. McDonald, "Soldier Settlement and Depression Settlement in the Forest Fringe of Saskatchewan," *Prairie Forum,* v. 6, n. 1, 1981, 35–56; T.J.D. Powell, "Northern Settlement, 1929–1935," *Saskatchewan History,* v. 30, n. 3, autumn 1977, 81–98.

tury and were part of a separate world unto themselves. In taking views of the fields, gardens, and meadows of the region, Crean was also trying to confirm the great expectations for the northland. The local inhabitants may have been growing cereals and vegetables to add variety to an otherwise monotonous diet, but from a different perspective, these efforts suggested that the region was destined to become a fine mixed-farming district. Together, the photographs constitute a wonderful visual portrait of a land and its people. They provide a revealing glimpse of a Northwest—whether new or old—that was little known to Canadians at the beginning of the twentieth century, let alone today.

The Photographs

The following 89 images have been selected from the Frank J.P. Crean collection at the Saskatoon office of the Saskatchewan Archives Board. In July 1978 the board purchased a photograph album and scrapbook of newspaper clippings from The Photographic Antiquarian (Canada) Limited, formerly of Montreal. The album contained a set of three-by-five, untitled, bromide photographic prints that had been pasted on black construction paper; the only link to the Crean expedition was the album cover, entitled "Photographs, New North-West Expedition, Frank J.P. Crean, C.E." The SAB secured a second Crean photograph album shortly thereafter, bringing the total number of prints to 305. This second set (with numbers directly on the prints) duplicated many of those in the first collection; it also contained a few scenes from an apparent third expedition to northeastern Alberta in 1910. The Crean photographs held by the SAB are evidently the most complete surviving set. The National Archives of Canada has a sampling of Crean photographs, but they have been integrated into the larger Natural Resources Intelligence Service collection (Accession 1936-271).

The photographs were taken by Crean or one of his assistants, William Caldwell in 1908 and Alfred Beale in 1909. Each expedition was equipped with standard Kodak cameras (No. 1A and No. 3A models) as well as a developing tank, packets of chemicals, and negative albums. Crean developed most of the exposed film in the field in the late fall while waiting for winter to set in; this may explain the poor quality of some of the surviving prints. The identification of the photographs was also a vexing problem. None of the pictures were labelled, and the SAB's identification of the images had to be based on a careful reading of the expedition reports and examination of related contemporary sources. This process was hindered by the fact that the 1908 and 1909 photographs were intermixed in the albums. In a few instances, the location or individuals remain unknown.

The Crean expedition left Prince Albert on 20 August 1908 and followed the Green Lake Trail northwestward through the recently settled Shellbrook district. (S-B 8792)

Crean was so impressed with the crops around Mont Nebo, sixty-five miles west of Prince Albert, that he sent a sample of wheat to Ottawa. (S-B 8794)

The expedition's outfit was transported by wagon to the Hudson's Bay Company freight depot at the south end of Green Lake. (S-B 8799)

Crean making notes while camped along the Green Lake Trail. His expedition notebooks have regrettably not been located. (S-B 8800)

Horses and wagons at the freight depot at the south end of Green Lake. The 150-mile trip from Prince Albert had taken five days—the last two through steady rain. (S-B 8854)

Transport boat and canoe on Green Lake, August 1908. The Prince Albert-Green Lake Trail had been opened by the Hudson's Bay Company in the mid-1870s to improve access to its posts along the Churchill River transportation corridor and further west. (S-B 8864)

The Green Lake settlement, late August 1908. Though local residents reported consistent success with their farming efforts, Crean was frustrated by their failure to keep records of seeding and harvesting dates and crop yields. (S-B 8882)

From left, Father Jules E. Teston, HBC clerk Edward Beatty and his family, and an unidentified Royal North-West Mounted Police constable at the Green Lake post. Teston served at Green Lake from 1890 to 1923 and from 1938 to 1941. (S-B 8858)

The children of HBC *clerk Edward Beatty at Green Lake.* (S-B 8861)

St. Julien's mission at Green Lake. Northcentral Saskatchewan had been the scene of an intense mid-nineteenth-century rivalry for Aboriginal souls, and the many missions established as a consequence provided focal points for Crean's explorations into the area. (S-B 8878)

Royal North-West Mounted Police constable walking through oat field at Green Lake mission. Crean advised Ottawa that the land around Green Lake was "first class" and that the country to the west was reportedly even better. (S-B 8863)

Crean, on horseback, explored east of Green Lake into the northwest corner of present-day Prince Albert National Park. (S-B 8874)

Father F.-X. Ancel welcomed the expedition to the new mission at the confluence of the Beaver and La Plonge rivers (now Beauval) in early September 1908. (S-B 8919)

The La Plonge wheat field appeared promising but had been damaged by an early frost. (S-B 8934)

Crean standing in a field of oats that reached his shoulders, La Plonge, 4 September 1908. This photograph was given prominence in the expedition's official report. (S-B 8938)

The mission sawmill at La Plonge. One of the first beneficiaries of the mill was a new residential school (left background). (S-B 8911)

The new three-storey residential school at La Plonge. At the time of Crean's visit, the school was nearly finished and soon to be occupied. (S-B 8916)

The Hudson's Bay Company post at Île-à-la-Crosse as seen from the waterfront, September 1908. Although both post and mission were renowned in fur trade and church circles, the site was not suitable for agriculture because of the low, swampy character of the land. (S-B 8941)

Local fur trade personnel and Royal North-West Mounted Police at the Hudson's Bay Company post, Île-à-la-Crosse. (S-B 8942)

Angus McKay, HBC *clerk-in-charge at Île-à-la-Crosse, served the company from 1877 to 1921. At the time of his retirement, he was post manager at Lac La Ronge.* (S-B 8940)

St. Jean Baptiste Church at Île-à-la-Crosse. The Oblate mission was founded in 1846, and over the next half century acquired a reputation as the "Bethlehem of the North," especially with the arrival of the Grey Nuns in 1860 and the building of a convent, orphanage, and school. (S-B 8963)

Revillon Frères post at Île-à-la-Crosse. The Paris-based company opened a district office in Prince Albert in 1901 and established a number of posts in direct competition with the HBC throughout northern Saskatchewan. (S-B 8970)

Two men outside the Revillon Frères post. (S-B 8971)

Chipewyan village at the mouth of the Buffalo River (now Dillon), September 1908, where Crean photographed a large vegetable plot and small patches of oats and barley, more proof to support his initial conviction that the north was "convertible into a fertile land." (S-B 8833)

Constable Thompson of Royal North-West Mounted Police at Buffalo River. (S-B 8836)

Frank Crean, with Kodak camera, taking photographs at Buffalo River. Crean developed all his exposed film en route, often in the most primitive of conditions, which may explain the poor quality of some of them. (S-B 8845)

Woman and children at Buffalo River. (S-B 8857)

Garden at Buffalo River, 12 September 1908. Crean was enthusiastic about the potential of the land northwest of Buffalo Lake (now Peter Pond). (S-B 8829)

Portage La Loche, September 1908. The post had served as the trans-shipment centre to the Mackenzie Basin until the route was abandoned in the mid-1880s. Here Crean learned that the region had once supported considerable mixed farming, but that much of this activity had come to an end when the route fell into disuse. (S-B 8883)

Portage La Loche served as temporary expedition head-quarters for several days. (S-B 8904)

The Hudson's Bay Company post, Portage La Loche. Priests from the local mission explained that although the land was good, their fields were small because of a lack of sufficient seed. (S-B 8893)

Mr. and Mrs. John O. Groat, HBC *clerk, and daughter, and an unidentified older man at Portage La Loche. Because of an apparent drinking problem, Groat served only one "outfit" year (1 June to 31 May) at La Loche before being released in 1909.* (S-B 8892)

Daughter of HBC *clerk John Groat, surrounded by the ubiquitous large dogs present at every point of human habitation in the north at the time.* (S-B 8884)

Constable Thompson of the Royal North-West Mounted Police on Methye Portage (or Portage La Loche), 1908. The gruelling twenty-kilometre portage, between the Hudson Bay and Arctic watersheds, had once served as the major fur trade route into the Athabasca country. (S-B 8923)

Valley of the Clearwater River, looking west from the heights at Portage La Loche (Methye Portage), 1908. Crean concluded that the region was comparable to the North Saskatchewan River valley and that one day it would be a magnificent cattle range. (S-B 8811)

Unidentified log buildings on a large lake in northwestern Saskatchewan, 1908. Crean visited the Canoe Lake mission (now Canoe Narrows), west of Île-à-la-Crosse, in early October. Although he found some good gardens at the mission, he decided little of the region was fit for farming because of the prevalence of muskeg. (S-B 8887)

Two of Crean's packers in camp, 1908. *Crean would have travelled more extensively in the region but found it difficult to secure guides who were willing to go inland, away from the well-travelled routes.* (S-B 8847)

Crean (left) and possibly his field assistant, W.R. Caldwell, sitting outside their tent, 1908. Due to the reluctance of guides to go off the beaten path, Crean and Caldwell often made little exploratory trips on their own. (S-B 8856)

Cutting oats by hand at Île-à-la-Crosse, 1 October 1908. The oats were likely used, in part, to feed the teams of horses that worked the Green Lake Trail. (S-B 8953)

Children outside the HBC *post at Île-à-la-Crosse,* 1908. *While waiting for freeze-up, Crean occupied his time by photographing local people, who proved willing subjects even when the camera was empty.* (S-B 8959)

Children, probably at Île-à-la-Crosse, 1908.
(S-B 8962)

Métis boy and girl at Île-à-la-Crosse, 1908.
(S-B 8968)

Family at Île-à-la-Crosse. (S-B 8880)

A group of packers travelling in early winter, 1908. Crean spent over two months at Île-à-la-Crosse, waiting until there was sufficient snow to continue his exploration eastward by dog team. (S-B 8851)

Dog teams at Buffalo Narrows. Crean and his assistant, Caldwell, travelled east to Lac La Ronge in late November 1908. (S-B 8835)

Stanley Mission on Lac La Ronge, early December 1908. At the time of the visit, there was considerable speculation about potential mineral deposits in the region, but Crean put more faith in the productivity of the soil, especially after learning that wheat at one time had been grown successfully enough to merit the construction of a mill at the mission. (S-B 8822)

Holy Trinity Church at Stanley Mission. Built in 1856, ten years after a Church of England mission was established in the Lac La Ronge area, it is the oldest existing structure in Saskatchewan. (S-B 8871)

Hudson's Bay Company clerk Angus McLean, his wife Lucie (nee Cook), and their family at Montreal Lake post, 7 December 1908. McLean served in the HBC Saskatchewan district from 1878 to 1921. (S-B 8808)

Shortly after his return to Prince Albert in early January 1909, Crean left for Nelson, B.C., to be married. At La Loche, Crean had arranged for the manager's wife to make some fancy work for his bride. (S-B 9092)

The Hudson's Bay Company headquarters and store at Prince Albert, June 1909. Less than three weeks after his report on the 1908 expedition had been submitted, Crean was instructed to continue his exploratory work into northeastern Alberta during the summer and fall of 1909. (S-B 8985)

Group of people outside the HBC *store at Prince Albert as outfit is loaded for second Crean expedition. Crean's 1908 report,* Northland Exploration, *had created quite an interest in the future of the region.* (S-B 8978)

Loading an expedition canoe for transport by wagon along the Green Lake Trail. (S-B 8984)

The expedition's outfit being readied for departure on River Street, Prince Albert, 17 June 1909. (S-B 8983)

The second Crean expedition leaving Prince Albert on 17 June 1909. Its mandate was the same as for the first: Observe everything and in so doing, confirm the great expectations for the region. (S-B 8981)

The North Saskatchewan River, upstream from Prince Albert. (S-B 8990)

Frank Crean (sitting foreground) with 1909 expedition members (possibly assistant A.M. Beale standing behind Crean). (S-B 9008)

Members of the second Crean expedition setting up base at Green Lake, late June 1909. The local HBC *post served as expedition headquarters for the first two months.* (S-B 9018)

Haying at Green Lake. Crean was greatly impressed by the fine hay lands he encountered exploring westward from Green Lake towards Lac des Îles. (S-B 8992)

The expedition members canoeing along the Beaver River. (S-B 8993)

Pushing canoes through a swamp, Waterhen Lake region, 1909. In some areas, the men spent more time outside their canoes than they did paddling them. (S-B 9050)

Crean and other expedition members at their Waterhen Lake camp, July 1909. (S-B 9042)

The expedition camp on Waterhen Lake. Although the land near the junction of the Beaver and Waterhen rivers was extremely swampy, it improved markedly as the expedition moved westwards. Healthy stands of hay and abundant undergrowth suggested the land must be good, even though it was not planted to crops at the time. (S-B 9033)

Frank Crean making observations with a sextant, Waterhen Lake region. (S-B 9039)

Two young Indian men in the Waterhen Lake area, 1909.
(S-B 9035)

Indian guides employed by Crean, Waterhen Lake camp.
(S-B 9041)

Drying fish in an Indian camp, Waterhen Lake area, 1909. (S-B 9036)

Frank Crean (far right) and the members of his second expedition. In addition to his departmental assistant, Alfred Beale, Crean was allowed to hire five extra men in Prince Albert. (S-B 9051)

Expedition members taking a break along the series of lakes that make up present-day Meadow Lake Provincial Park. (S-B 9044)

A heavily laden Crean, July 1909. Crean was a rough-and-tumble character whose weakness for booze was his eventual undoing. (S-B 9052)

T.G. Street portaging expedition equipment south from Lac des Îles to the east-west arm of the Beaver River. Since a survey team was already working in the Cold Lake area west of Lac des Îles laying out the Saskatchewan-Alberta border, the expedition decided to head south towards Meadow Lake. (S-B 9049)

A wagon and team of horses going through tall red-top grass, Meadow Lake, August 1909. (S-B 9021)

A field of Banner oats at the farm instructor's house, Meadow Lake Reserve, 1909. Crean would later declare that the district contained "some of the very finest farm land in Canada." (S-B 9029)

A settler's family in their garden at Meadow Lake. (S-B 9032)

The frame of an Indian sundance lodge in the Meadow Lake district.
(S-B 9025)

Two young men at Île-à-la-Crosse, August 1909. After completing his assessment of the land in the Meadow Lake district, Crean moved north to Île-à-la-Crosse and then on to La Loche, where he established a new base of operations. (S-B 9054)

Sweet peas in the flower garden at the Hudson's Bay Company post, Île-à-la-Crosse. (S-B 9055)

Potato field along the road to Whitefish (now Garson) Lake, late August 1909. Except for some small plots of root vegetables, land around the lake was generally swampy and unsuitable for crops. (S-B 9066)

Crean (rear) and one of his assistants ascending rapids on the Whitefish River. The party's decision to approach Whitefish (now Garson) Lake by canoe instead of the more direct summer cart trail was quickly regretted, since the boats had to be dragged for miles along the shallow, snag-ridden Whitefish (Garson) River. (S-B 9083)

Unidentified settlement in northwestern Saskatchewan (possibly present-day Garson Lake), 1909. (S-B 9077)

The expedition travelled down the Clearwater River to Fort McMurray, September 1909. Crean was enthusiastic about the agricultural possibilities of the river valley, declaring that nature had been "extremely bountiful" to the region. (S-B 9082)

Miss Christina Gordon and her wheat field at Fort McMurray, September 1909. Any lingering doubts Crean may have had about the valley's potential were easily put to rest by the vision of this healthy vigorous stand of grain. (S-B 9097)

Men and dogs in a camp in northeastern Alberta (possibly around Cowper Lake) south of the Clearwater. Growing conditions around Cowper Lake were so good that local Indians sowed gardens in the spring and did not return again until the fall to harvest the produce. (S-B 9058)

Cutting planks from a log, northwestern Saskatchewan (possibly La Loche Lake), 1909. (S-B 9061)

An expedition canoe, equipped with sail, probably on La Loche Lake, 1909. (S-B 9071)

The expedition base camp at Portage La Loche, 1909. While waiting for freeze-up, Crean visited some of the larger lakes in the region and witnessed the annual fall fisheries. On one day alone, 12,000 fish were hung at the north end of Buffalo (Peter Pond) Lake. (S-B 9075)

T.G. Street and Royal North-West Mounted Police Constable Thompson beside an old wagon on the Portage La Loche (Methye Portage) Trail, 1909. (S-B 9084)

In the late 1840s, Sir John Richardson travelled through northern Saskatchewan while searching for Sir John Franklin. This man, living in northeastern Saskatchewan in 1909, had assisted Richardson some sixty years earlier. (S-B 9087)

Frank Crean (centre) and his winter travelling companions, November 1909. From Portage La Loche, Crean travelled west to Fort McMurray and then south to Edmonton. (S-B 9104)

Frank Crean taking measurements with a sextant while travelling through northeastern Alberta, November 1909. The New Northwest expeditions marked the zenith of his career and he resigned his position with The Department of the Interior in 1913. (S-B 9106)

Note on Sources

There is little surviving material from the New Northwest explorations. Most of the federal administrative records associated with the Railway and Swamp Lands Branch in general and Robert Young's northern promotional work in particular were destroyed in the 1930s. As a consequence, the National Archives of Canada has only three slim files (Interior records) dealing with the Crean expeditions: an itemized list of the 1908 expedition outfit, a typescript of the expedition report, and the interdepartmental correspondence regarding Crean's 1913 resignation. There is also no mention of the Crean expeditions in the annual published reports of the Department of the Interior—a puzzling omission given the national press coverage at the time. Fortunately, the Crean material acquired by the Saskatchewan Archives Board in 1978 included an indexed scrapbook of newspaper articles on the region (possibly Robert Young's), as well as copies of the 1908 Crean–Young correspondence, including that while Crean was in the field.

The best sources for this study were the various publications released by the Department of the Interior under Robert Young's supervision. The 1907 Senate hearings on the middle north (the Davis committee) appeared in pamphlet format the following year: E.J. Chambers, ed., *Canada's Fertile Northland: A Glimpse of the Enormous Resources of Part of the Unexplored Regions of the Dominion* (Ottawa 1908). Chambers also prepared a condensed version of the 1887 and 1888 Senate hearings on the region (the Schultz committees) entitled *The Great Mackenzie Basin* (Ottawa 1908). Frank Crean's two expedition reports were published together under the title *New Northwest Exploration* (Ottawa 1910). Finally, an edited compilation of all the available information on the resources of the Northwest appeared in 1914: E.J. Chambers, ed., *The Unexploited West* (Ottawa 1914). There is little if any mention of Young's activities or Crean's explorations in the existing secondary literature.